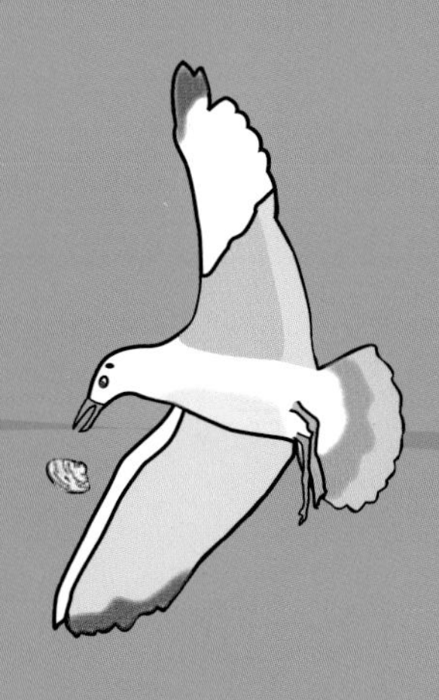
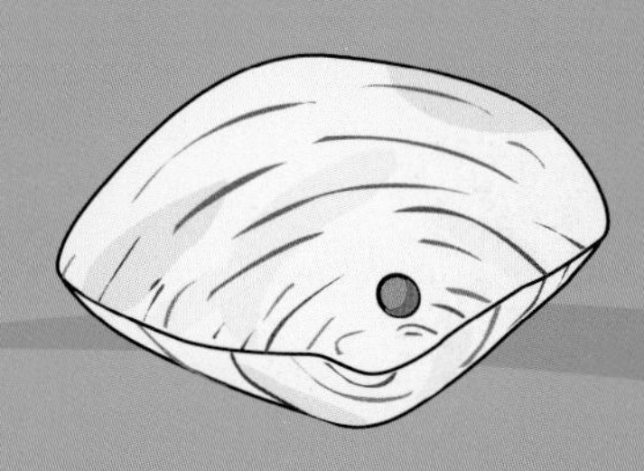

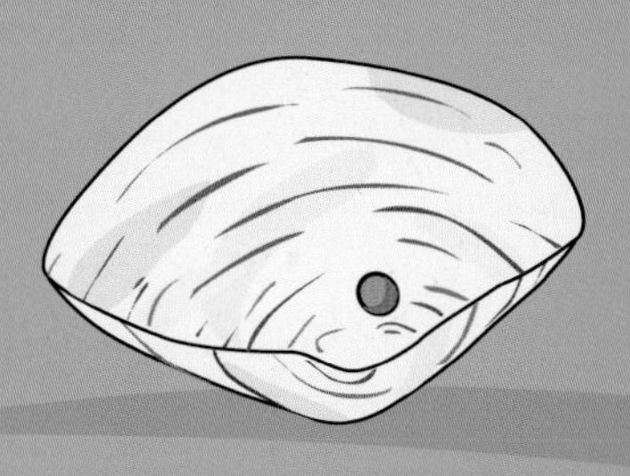

ACKNOWLEDGMENTS

To Mike and my family:
Here's to soaring with the eagles!

First Printing: October, 2013

 Printed in the United States of America.

ISBN 10: 0-9705510-2-9
ISBN 13: 978-0-9705510-2-3

Did You Make the Hole in the Shell in the Sea?

Written and Illustrated by

Janice S. C. Petrie

Holy Mackerel,
what could it be?
There's a hole in a clamshell
under the sea!
Ooooooh!

Oh Mister, hey Mister,
can you tell me?
What made this hole
in the shell in the sea?

Let me see, oh dear me!
Can it possibly be?
A shark made this hole
in the shell in the sea!

A shark, it's a shark,
whose tooth bit right through it.
I knew when I saw it,
a shark's tooth could do it!

Get out of the water,
hurry up, make it snappy.
There's a shark on the loose,
and he isn't too happy!

Are you sure it's a shark?
No one's seen one around.
Sharks make MANY holes,
not ONE like we found.

I need to look further
to know what to do.
Some other sea creatures
might know what is true.
Oh sea star, hey sea star,
can you tell me?
Did you make the hole
in the shell in the sea?

Not me, not me,
I must disagree!
I love to eat clams,
but it wasn't me!

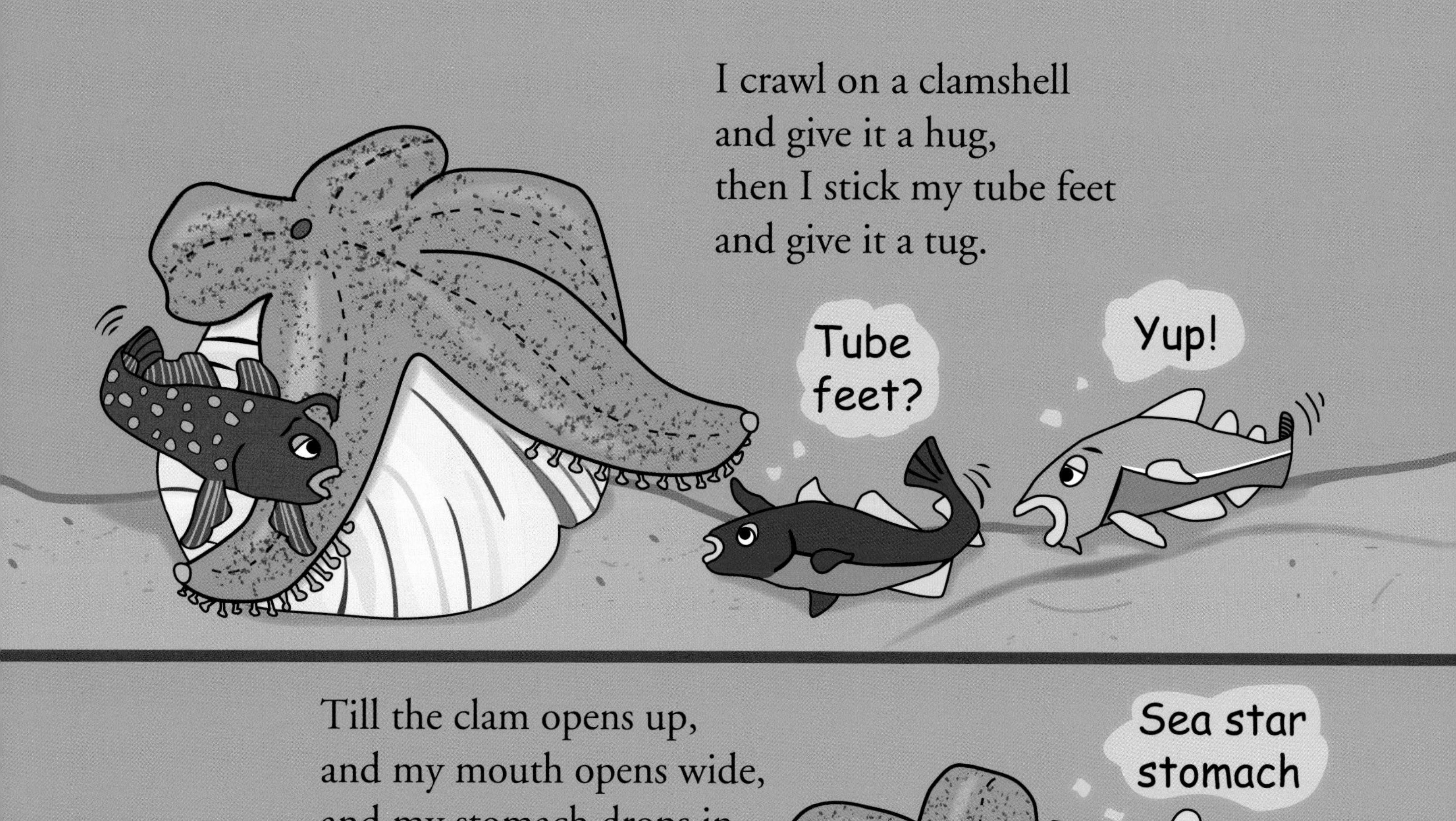
I crawl on a clamshell
and give it a hug,
then I stick my tube feet
and give it a tug.
Tube feet?
Yup!

Till the clam opens up,
and my mouth opens wide,
and my stomach drops in,
and I slurp what's inside.
Sea star stomach
Ewwww!
Cool!

Anybody home?
Nope
And when I'm done eating,
and I crawl away,
the clamshell is perfect
in every way.
Just perfect!

Oh seagull, hey seagull,
can you tell me?
Did you make the hole
in the shell in the sea?
Not me, not me!
It's plain to see.
I love to eat clams,
but it wasn't me!

When I see a clam
I scurry to catch it,
so another seagull
isn't able to snatch it.

I puff my chest up,
and I let out a cry,
and I spread out my wings,
and I take to the sky.

I drop the clam hard
to shatter its shell.

Then I swoop down to eat
from the rock where it fell.

Oh lobster, hey lobster,
can you tell me?
Did you make the hole
in the shell in the sea?

Not me, not me,
I guarantee.
I love to eat clams,
but it wasn't me!

When I'm crawling around
and happen to spot,
a sweet, juicy clam
that hasn't been caught.

I circle the clam
with my claws in the air,
keeping others away
who think I might share.

Then my crusher and scissor claws
crack and they crunch,
till the clamshell's in pieces
and the clam is my lunch.

Oh moon snail, hey moon snail,
can you tell me?
Did you make the hole
in the shell in the sea?

Who me, who me?
Quite possibly.
I love to eat clams,
and I bet it was me!

I climb on a clam
with my smooth, slimy foot.
And hug the clam's shell,
to make sure I stay put.

I soak the clam's shell
with a shell-softening goo.
And the teeth on my tongue
scrape a hole right on through.

Then my tongue reaches in,
and scoops up what's inside.
And when I'm done eating,
away I will glide.

And as I am leaving
you plainly can see,
a precisely drilled hole
in a shell in the sea.

Oh Mister, hey Mister,
the mystery's solved!
There never was ever
a shark here involved.

It's a moon snail, a moon snail,
whose tongue drilled right through it.
I saw it myself.
Who knew he could do it?

The danger has passed,
and it's hotter than hot.
Let's go back in the water.
The culprit's been caught.

Meet the sea animals:

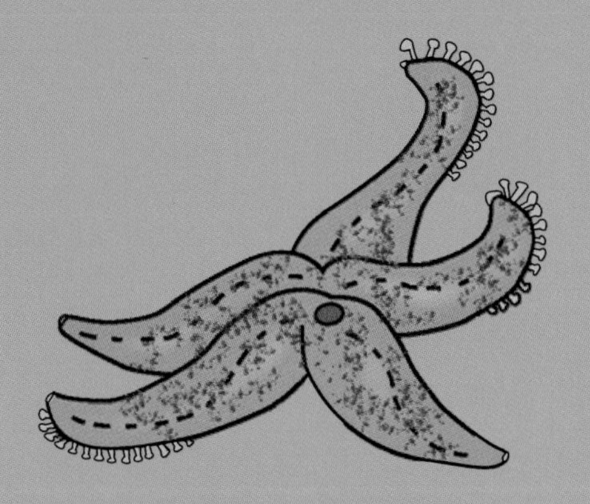

Scientists changed the name of the starfish to **sea star** because sea stars aren't fish, they're echinoderms. The common sea star has five rays, with an eyespot on the end of each, which can sense shadows and light. The reddish-orange, round spot on top of the sea star is called a sieve plate or a madreporite. The sea star brings water into its body through its sieve plate. The water flows through a system of tubes to help the sea star move, eat, breath, and cling to things. On the bottom of each ray, the sea star has many tube feet with a suction cup at the end of each one. These tube feet help the sea star to move and to hold tightly to rocks, and especially shellfish, which is the sea star's favorite food. A sea star tugs on a clam with its tube feet to part its two shells. Then the sea star opens its mouth, which is in the center of its body underneath, and drops its stomach out of its mouth into the clamshells to eat the soft clam body inside. When done eating, the sea star pulls its stomach back in through its mouth. To this day, it's the only animal I've ever known that likes to take its stomach out to lunch!

A **moon snail** is a mollusk that has a hard shell for protection, and a hard trap door called an operculum, to cover the opening of its shell. The moon snail's huge, smooth, slimy foot can practically cover its shell. This foot allows the moon snail to gracefully glide over the sand. The moon snail can make its foot larger by filling it full of seawater. When the moon snail needs to hide in its shell, it removes the seawater from its foot to make it shrink in size. Moon snails eat clams and other bivalves, which can be found under the sand. A moon snail digs and travels under the sand using its foot, in order to hide from predators and to hunt. Once it finds a clam, the moon snail surrounds the clam with its foot, and pulls out its tongue called a radula. Its radula has teeth that work like a chain saw, rapidly scraping forward and back until it's drilled a perfect hole in the clamshell. The moon snail releases an acidic enzyme while it drills, to soften the shell and make the drilling easier. Once the hole is drilled, the radula reaches inside the clam's shells to eat the soft body. When it crawls away, the moon snail leaves behind a precisely drilled hole in the shell in the sea.

Seagulls are true scavengers and will eat almost anything. They gather in groups called colonies for protection, and fight with each other for food. When a seagull finds a clam, it flies high until it finds an updraft to glide on. Then the seagull drops the clam on a rock below to break open its shells. Quickly, the sea gull swoops down to eat the clam before another seagull can steal it.

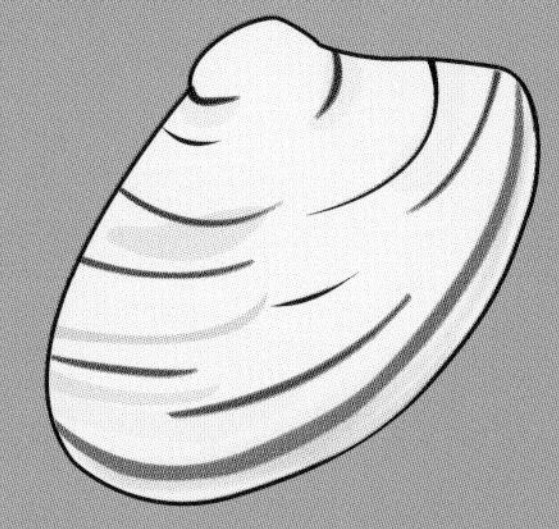

Clams are mollusks that have a soft body enclosed in two shells held together by a hinge. Because a clam has two shells, it's called a bivalve. A clam's mantle is responsible for making the clam's shells, and the mantle thickens at the back to form two siphon tubes. Clams are filter feeders that take in water through one siphon tube, and after the food and oxygen are removed; the remaining water is released back into the ocean through the second siphon tube. Clams have a fleshy foot that reaches outside the clam's shells, which helps the clam to move and dig under the sand. People and lots of sea animals love to eat clams.

A **lobster** is an arthropod, because it has a segmented exoskeleton and jointed legs. Its two eyes, which are mounted on stalks, are compound eyes, which are able to detect movement extremely well. A lobster has two long antennae that it uses to feel its way. Lobsters have two shorter sets of antennae that can sense odors and distinguish foods. A lobster uses its sharp scissor claw to cut its food and its strong crusher claw to hold and crack open its food. These powerful claws are also used to protect the lobster. If a lobster wants to escape from an enemy, it flaps its tail rapidly to swim backwards and hide under a rock or a branch of seaweed. Underneath the lobster's tail or abdomen, the lobster has many swimmerets that it uses in combination with its legs to crawl along. If the lobster's first set of swimmerets is hard, the lobster is a male. If the first set of swimmerets is soft like a feather, the lobster is female. Although lobsters are considered scavengers, they prefer to eat live foods like fish, plants, and mollusks, including clams.

Shark sightings have increased as the seal population has exploded in recent years. Sharks, for the most part, are carnivores, or meat eaters, which have rows of very, pointy teeth. If a shark loses a tooth, there's another one right behind it to take its place. A shark has a keen sense of smell, and can sense movement with its lateral line that runs from its head to its tail. By being able to feel vibrations in the water, the shark easily finds its prey. The great white shark has a white belly, which from below looks like the sky. Its gray color on top helps the shark to blend in with the ocean floor. Although statistics vary, according to National Geographic, there are about nineteen shark attacks in the U.S. each year, with one death happening about every two years.